A BEGINNER'S GUIDE
TO CREATION

A BEGINNER'S GUIDE TO CREATION

Chris Pegington

BRYNTIRION PRESS

ISBN 978 1 85049 229 0

Cover design: Creative Media Ltd

Published by Bryntirion Press
Bryntirion, Bridgend, CF31 4DX, Wales, UK
Printed by Gomer Press, Llandysul, Ceredigion, SA44 4JL

Contents

Foreword

WE live in days when there is a battle for the mind of the people and especially our children. Of course, this has always been the case from the time of the Fall (Genesis 3) and the underlying issues are the same. In recent years this has become focused on the issues much publicised by Professor Richard Dawkins and his allies.

The issue is ultimately, 'Who is God?' The figment of Dawkins' imagination does not correspond to the God of Scripture, the one true God. The argument of such people is that 'Science says . . .', and it is placed in contradiction to 'The Bible says . . .'. Science is elevated to a position of authority that it does not have, but people are made to feel ignorant and foolish if they believe the Bible rather than the edicts and theories of some famous scientist. Nowhere is this more evident than in the matter of our origins. Do we believe the Bible or some scientists?

This has even affected evangelicalism, in which there are many who accept the word of scientists rather than the obvious teaching of Scripture. It is into this conflict that Dr Chris Pegington has entered the fray with this very helpful publication. Dr Pegington is qualified to write on this from his biological background and his biblical experience. He has spoken frequently on the issue, and the style of the book reflects this. He communicates simply and clearly for the non-specialist reader and yet with the authority that his studies have given him.

In some areas there will be debate amongst creationists on how to interpret the details of the biblical account (for example, in geology), but such issues do not deflect from the usefulness of this publication. I expect Christians to read this book and be encouraged in the reliability of God's Word. Non-Christians can also read it with profit and see that we have a faith that stands up to the tests of the twenty-first century as well as those of the first century.

We are in debt to Chris for this publication and I know the Lord will be pleased to bless it.

J. H. John Peet
Travelling Secretary, Biblical Creation Society, retired lecturer in Chemistry, Ph.D., F.R.S.C., C.Chem, author of *In the beginning GOD created* . . . (Grace Publications Trust) and co-author with John Benton of *God's Riches* (Banner of Truth Trust).

Introduction

WHEN I left school and started at university in 1963, I was a convinced evolutionist. The A level Biology course I had completed not only assumed that evolution was true, but stated it as a fact. We were simply expected to learn the proofs of evolution for the examination. No one even suggested that there might be other explanations for the origin and development of life on earth; the only view we were presented with was that of evolutionary theory.

It was believed that life had arisen by chance in the distant past and that those early life forms had gradually changed into ever more complex forms. In some form of chemical soup, the first single-cell forms of life had appeared spontaneously on earth several billion years ago, to be followed by the multitude of multi-cellular forms over the millions of years that have followed. Among the animals this could be summarised as the FARM series: Fish gave rise to Amphibians, from which developed the Reptiles, and later the Mammals evolved. This process of development from Amoeba to Man was held to be well-documented, and no serious scientist believed otherwise.

By the time I left university life, nine years later, after completing a doctorate in genetics and being involved in research and lecturing at the University of Cambridge, I was a convinced creationist! This book is a brief attempt, aimed at the general reader with little training or background in science, to explain, first, how that came about and, secondly, to show how a belief in the teaching of the

Bible is compatible with the facts that scientific research has revealed.

In the early 1960s there was relatively little material that called into question or challenged the evolutionary view of origins accepted by the overwhelming majority of scientists. Since then, the growth of creationist literature and influence has been phenomenal. Surveys in the USA have shown that over 50 per cent of Americans now believe that the Bible's account of creation is true rather than the classic evolutionary model. The percentage in the UK is likely to be much smaller, but even here the number of scientists and non-scientists who have doubts about evolution—both Christian and those of other beliefs or none—is not insignificant. The reason for that change is mainly due to the influence of creationist organisations and literature, although new discoveries in biology and other scientific disciplines have also played their part.

I have sought in this book to avoid technical language as far as possible, so that it can be understood by the non-scientist. For those readers who wish a more rigorous and detailed refutation of evolution and to examine the creationist alternative in more depth, there are plenty of good books and websites available. A small bibliography is attached at the end of the book.

1
General and Specific
(or limited) Evolution

When is a fact not a fact? When it is an interpretation!

B EFORE we discuss the topic of evolution, there are a few basic things about which we need to be clear. First, what do we mean by evolution? The term is often used in different ways by different people. There are really two main theories of evolution. The first is the *General Theory of Evolution*, to which I alluded in the Introduction. That is the theory (often presented as, or assumed to be, proved) that all life came originally from inanimate matter by chance, and that original life, in whatever form it appeared, has gradually changed over the years into the species that now exist, or have ever existed, on this planet.

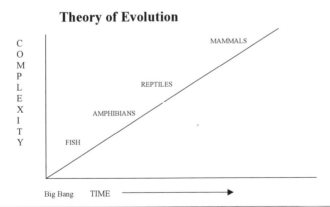

Theory of Evolution

Over millions of years organisms develop into ever more complex organisms through the acquisition of new genetic information. The changes are widespread, slow and small and cumulative. At the same time the variety of organisms, represented by the area under the sloping line, also increases.

There is also, however, what we can call the *Special Theory of Evolution*, which describes how different plants and animals can change and develop over time into slightly different forms of the same basic type or kind. The mechanisms causing these changes in varieties fall largely within the realm of genetics and are sometimes referred to as *speciation*. That some changes can occur in the appearance of living species has been frequently demonstrated. An everyday example is the domestic dog, which ranges in shape from a poodle to a Great Dane, from a greyhound to a Yorkshire terrier. This great variety of forms has been developed by human selective breeding from just a couple of wild dog species over a few thousand years. Similar examples could be given of many animals and plants that have shown dramatic changes over historical time.

Creation of the Kinds

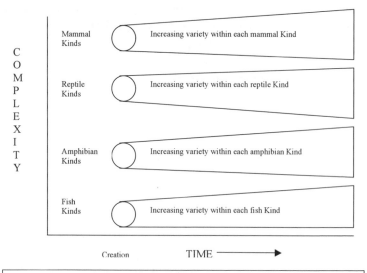

God created in the beginning different kinds of plants and animals. Over time a greater variety of organism within each kind has appeared (and disappeared), but the boundary of each kind of plant and animal has remained intact.

Now an important question to ask is whether the biological processes that give rise to these small and limited changes within dogs, for example, can also explain the major changes that have given rise to the dog kind itself. We will seek to answer that question in chapter 4. From now on, in order to avoid confusion, whenever I use the term evolution, I am referring to it in its general sense.

Some assumptions of evolution

A second question we need to be clear about is, What is a fact? A fact can simply be an observation that is normally true: for instance, that 'cows have four legs'. Sometimes, however, what we perceive of or is even claimed as a scientific fact is actually just an interpretation of certain observations that have been made. Later, perhaps, those interpretations are verified by further observations and so obtain the status of being proved; sometimes they either remain unverifiable or, being contradicted by later studies and experiments, are rejected or subsequently amended. This is the normal scientific method by which our understanding of the physical and biological world advances. However, whenever we begin to interpret an observation or experiment, the assumptions we make before we carry out that exercise become important.

One of the fundamental underlying assumptions of evolution is *naturalism*—that all of this universe, including the origin and development of life, can be explained by natural processes, without the intervention of or need for a divine or supernatural being. A second assumption is that we can explain what has happened in the past on the basis of what is happening in the present. This belief, sometimes explained as 'The present is the key to the past' (*uniformitarianism*), is clearly unprovable, as no one can

13

go back into the past to check it out. We will also see in due course that it is an unreasonable assumption. Both assumptions, however, remain accepted as valid by most of those involved in evolutionary research, and some scientists insist that naturalism is essential to the scientific method. Any assumption of a divine power being involved in creation of the universe and the origin of life cannot be regarded as science at all, they claim. This prior commitment to atheistic naturalism partly explains the vehemence of some of the opposition to those who espouse creationist alternatives to evolution. It is not that the creationist position has no observational support, but rather that it is ruled out *a priori* on philosophical grounds.

New assumptions
The beginning of my own questioning of evolutionary theory began, not with startling new scientific observations that compelled me to change my views of life and its origins, but with a new set of underlying assumptions. At the age of nineteen, when I was a first-year student at university, Jesus Christ changed my life. Hearing one day the gospel's message that Christ had died for my sins and was alive from the dead, I suddenly knew that it was true and I committed my life to him as Saviour and Lord. That decision changed my life, and with that change came a conviction that the Bible was indeed the Word of God, as it claimed to be. God had spoken to man through his infallible Word and he could be trusted to tell us the truth. I began to read the Bible's account of creation and the history of the world, and it began to raise questions in my mind about what I had hitherto believed about the origin of life.

Two books helped me enormously in thinking through these issues. The first was *The Twilight of Evolution* by

Henry Morris, and the second *The Genesis Flood* by Morris and Whitcomb. Here were two books where the implications of taking God's Word as true were worked out in relation to the natural world. Both books are now somewhat dated and much more work has been done by a host of other writers and researchers, but they opened for me the possibility of other explanations beside naturalistic theories for the origin of life.

Comparative morphology

Anyone who has studied the natural world at all cannot fail be impressed by the astonishing variety and complexity of the different plants and animals, both living and extinct. It is possible to arrange this wonderful variety into a graded series, from relatively simple forms of life right up to the most complex sorts. If we compare creatures at any one stage—the land mammals, for example—we find remarkable similarities between them. Every land mammal has two eyes, four limbs, one heart, two kidneys, etc. The same can be done for fishes, birds and reptiles.

Comparative Morphology

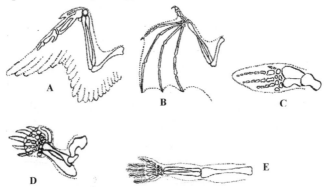

Comparative morphology is the study of the similarities and differences between similar parts of different creatures. In this diagram the forelimbs of various animals are compared. The obvious similarity in the number and arrangement of the bones is easily seen.

A: Wing of a bird. B: Wing of a bat. C: Dolphin flipper. D: Mole forelimb. E: Human arm.

Classic evolutionary theory has pointed to these observable likenesses as an evidence for evolution, or indeed as a proof of evolution. The mammals have the same basic ground plan, we are told, because they are all descended from the same original population of the first mammalian creatures to appear on earth. Just as, when we see two people who are very alike, we can reasonably assume they share the same parents or ancestors, so, it is claimed, similar morphology among the land mammals is a proof of common evolutionary descent and ancestry. School textbooks often have pages of comparisons between similar features, such as the forelimbs or the blood circulatory system of different animals. Here, we are told, is proof of evolution.

Books galore

Let us look at this argument for a moment. Suppose you went into your local town library. You would find a great variety of books on all sorts of subjects. You could, if you have the time, arrange these books into a long chain of books, starting with the simplest children's books right through to the most complex reference works. The children's books would have pages with some words and pictures. Those words would be made up of letters, and there would be examples of punctuation and grammar, with perhaps some numbers. As we examine the books further along our artificial chain, we would find other similarities as well as novel features (no pun intended!). The research books would have chapters and paragraphs, subheadings and footnotes; probably an index and bibliography, and so on.

Now we would be very foolish to argue, on the basis of our organising the books into this system, that the more complex books had somehow developed by chance from the first few children's books on our list. If we ask why the different books are similar in some ways and different in others, the answer would be quite simple. They were designed in that way by the publishers to meet the need of the end-users—the readers. The differences would reflect the different demands of those readers. Some of the similarities are due to the basic needs of any book that seeks to convey a message, whether to children or adults, artists or scientists. They are made to be that way in order to meet the varied demands that the books would have to meet. Historically, of course, children's books appear later than other types of literature, and certainly didn't give rise to them!

Looking again
Let us now see how this applies to the argument from comparative morphology. The similarities between different types of creatures arise partly from the necessity of life itself. At the cellular level, each organism needs mechanisms to produce energy, reproduce, repair itself, transmit information, etc. Looking again at the groups of land mammals, they all share similar challenges to survive. They need to move in order to eat; to breathe and digest in order to be able to oxidise their food to produce energy; to reproduce or they would quickly disappear from the earth. To move, they need to see where they are going; they need to hear in order to communicate; they require some way to defend themselves, and they have a host of other needs. A wise Creator, who designed them to live on the earth, made them so that they were capable of surviving. That he used the same basic pattern in order to enable them to do so shows it is the activity of one God, and not a multitude of gods; and he used a certain economy in creation by

employing the same basic blueprints in each case. A God for whom all things are possible could naturally have used a different basic pattern for every type of animal, but he did not. A further reason for the similarities may be that the basic design of each kind of creature is the best one possible for them to survive and flourish in the land environment.

Summary

Does the similarity of different animals prove evolution? Not at all; but neither does it prove that God has created them. The conclusions we arrive at, and the interpretation of the evidence available to us to study, will be different depending on the prior assumptions we bring to the study. However (and we will consider this in chapter 3), when we come to look at 'Intelligent Design' as a fairly recent scientific idea, we shall see that it does point to compelling evidence that there is an intelligent mind behind life on earth. And if God has created all life, as the Bible teaches, that is precisely what we would expect.

2
A waste of space?

WHEN I was thirty-seven years old I went into hospital and had my appendix removed. Two years later a similar visit to a different hospital involved the removal of my tonsils. Both operations were necessary because of disease in those particular organs. Since then I have suffered no ill effects from their absence. Evolutionary scientists tell us that this is because these organs, and others like them, have no present use in the body, so we do not suffer loss if they are taken away. They were once useful in our evolutionary past, but they have become redundant—mere vestiges, leftovers, in bodies that no longer need them.

These vestigial organs, as they are called, are paraded before us as proof that we are descended from the apes and other creatures. Among other parts of the human body claimed to be vestigial are: the *semi plica lunaris* (the pink flap of skin in the corner of each eye near the nose); the coccyx (the bony projection of the end of the spine caused by the fusing of the last four vertebrae); the pineal gland, embedded in the centre of the brain; the outer ear muscles, and many more. All these are proof of the Theory of Evolution, we are told.

Animal remains?
The *semi plica lunaris* is claimed to be the remains of the nictitating membrane—a transparent skin that sometimes covers the eye of some reptiles such as the chameleon. The

19

appendix is important in some herbivores, like the rabbit, where it is much bigger and plays an important role in the digestion of cellulose, which makes up the hard walls of the cells of plants. It is claimed that as our diet changed, so it became unnecessary and shrank to its present size. Likewise, we are told, our coccyx points to a time when we had tails, like some monkeys, but which we subsequently lost as we evolved further. Humans have well-developed ear muscles, despite being generally unable to move our ears—it is a good party trick if you can! The pineal gland is considered to be the remnant of the 'third eye', a light-sensitive organ that some reptiles possessed —and so on. I have not yet found anyone who can tell me of what the tonsils are supposed to be the evolutionary remains! Such vestigial organs have been claimed, moreover, to have been found in many species, not just in humans.

Why are they there?
Now if God created the different kinds of animals in the beginning, as the Bible seems to teach, we would not expect to find any such vestigial organs. So are vestigial organs real or not? Let us look at the argument a little more closely. We need to begin by asking, What is the logic underlying the vestigial organ claim? Basically it comes down to this. As modern science knows of no use for a particular organ in the human body, it is therefore useless! In other words, ignorance of a function is offered as proof that no such function exists! Having established that as a 'fact', the assumption that evolution has occurred is then used to explain that conclusion!

Now suppose you ask me to take a look under the bonnet of my car. It is a fairly modern car, with the engine space packed with different parts. You could ask me to name the various parts and tell you what they are there for. I would be able to identify the function of many parts and hazard a good guess at others, but about some parts I would be totally in the dark. However, if I went on to suggest that the fact that I didn't know the function of certain bits was proof that they were a waste of space and functionless, you would rightly consider me very foolish. Even when the removal of some parts seemed to have no immediate or obvious effect on the running of the car, it would be arrogance on my part and ridiculous logic to suggest that I knew better, with my limited knowledge, than the maker and designer of the vehicle. The parts under the bonnet are there because they are necessary.

Historical evidence

Looking back to the beginning of the last century, we find that Ernst Wiedersheim, a noted German physiologist, listed over 180 such vestigial organs in the human body. His list included many of the hormonal glands, which we now know are so important to the right functioning of the body. As our understanding of human physiology has advanced, that list has become smaller and smaller. Even for those organs about which there are still some questions as to their exact function, various possibilities have been suggested. The appendix and the tonsils, for example, seem to have a role to play in the body's defences against disease, particularly in babyhood and early childhood. What is useful to a person in helping him survive at that early stage of life cannot be said to be useless, just because its role is less prominent later in life.

In addition, studies show that people who have their tonsils removed are more likely to develop infections in the respiratory tract, and four times more likely to develop Hodgkin's disease; while the appendix is part of the gastro-intestinal immune system. There is evidence that the pineal gland produces melatonin and is involved in the regulation of circadian rhythms, and possibly in regulating the development of sexuality in children by inhibiting the development of sexual organs. The coccyx has important muscles attached to it and helps support the organs in the pelvic region, so that they don't fall down through our pelvis when we stand up! For the other organs on the list (and others I have not mentioned) various functions have been suggested. I am quite sure that as research continues and our understanding of our complex and amazing bodies grows, what today are so confidently called useless vestiges of our animal past will be found to be important parts of the human body. The same conclusion applies to other so-called vestigial organs in other animals.

Shifting goalposts
Before we leave this subject, however, it needs to be mentioned that evolutionists are not simply abandoning alleged vestigial organs as a 'proof' of evolution. We are now told that even where a function exists for an organ in an animal today, it can still be a vestigial organ! If, on the basis of supposed evolutionary descent, an organ appears to have become smaller, or to have a changed or reduced function, it is still, we are told, an indication of evolutionary redundancy.

Two things need to be said about this claim. First, the basis for this argument is once again obviously the prior commitment to the theory of evolution by chance. The assumption that evolution has occurred is used to explain certain observations and comparisons. That explanation is then used as a proof of that assumption—another clear example of circular reasoning.

Secondly, it is just too broad an argument to be meaningful at all. Nearly every part of the human body can be shown to be smaller than in some supposed ancestor of ours. Our jawbones, for example, are smaller than many of the postulated links in the human evolutionary chain. They certainly have a function in our bodies, but does their smaller size mean that they are vestigial organs? Of course not!

Summary

The claim of some evolutionists that we have redundant parts in our bodies that are indications of our evolutionary ancestry is found, on examination, to be baseless. It should be rejected for logical and historical reasons. Our increasing knowledge of our physiology and metabolism should cause us to cry out with the psalmist that we are 'fearfully and wonderfully made' (Psalm 139:14).

To furnish a proof of evolution, we need to see not vestigial organs, but newly forming organs that will confer some sort of advantage on the organism concerned. No such claims have been made, however, nor would we expect to see such if creation is true. There is no evidence of 'novel' organs appearing in any plants or animals. As we shall see in the chapter on fossils, fossil specimens of 'older' versions of modern-day animals and plants are scarcely different from their modern-day equivalents.

3
Intelligent Design
or blind chance?

In the previous chapter, when we looked at the similarities that exist between different land mammals, I suggested that these arise through the deliberate design of God, the Maker of the universe. Animals were created with those features they would need in order to survive in different environments on the land and in the sea. Birds, for example, have special features to enable them to fly; likewise fish to survive in water, land mammals on the land, whales in the sea.

Throughout history the observation that, everywhere we look, we seem to see evidence of perfect design features, has often been used as an argument for the existence of God. It found its classic expression in William Paley's illustration of the divine clockmaker in his book, *Natural Theology: or, Evidences of the Existence and Attributes of the Deity, Collected from the Appearances of Nature,* published in 1802. He argued that if we were to find a watch while out walking and examined its workings, we would conclude that it could not have just arrived there by blind chance, but there must have been an intelligent watchmaker who had made it. Atheistic philosophers have, not unexpectedly, dismissed his arguments out of personal necessity, since to admit their validity would add pressure towards a belief in a god or divine being they had denied already.

Opponents have argued, then, that this appearance of design is something we simply read into our observation of nature. Our desire for order and our urge for an explanation and reason for all things compel us to see patterns that are only accidental and have arisen through chance. There is no mind behind them, but our minds think we see one. Given enough time, we are told, anything is possible. The universe is so old that eventually even the seemingly impossible is bound to occur. Our faith in God misleads us into seeing things that just don't exist.

Before we continue, we need to remember that the rejection of a personal mind (God) behind creation, and attributing all things to chance, involves such a tremendous belief in the power of chance that it can only be categorised as being as much 'a faith' as any belief in a personal, wise, all-powerful, active God. Also, despite all his protestations to the contrary, the atheist remains as much a prisoner of his own atheistic assumptions as any Bible-believer is influenced by his own belief in God. The rise of 'Intelligent Design' theories is all the more remarkable, therefore, since many of those who support it are not Bible-believers at all.

What is it?
A number of biologists have noted that certain systems in nature possess what is called an irreducible complexity. That is, the system is so complex that it needs many separate and interrelated parts for it to work at all. If any bit of the system is missing or falls out, the system fails to function. At the same time the individual parts, on their own, are not capable of being of benefit to the organism involved. Only when they are combined with the other parts do they confer any benefit on the creature or plant possessing them. They give every appearance of being

designed to do the job that they do. It is also very difficult, if not impossible, to see how such systems could have arisen gradually over millions of years in small steps through pure chance.

Think for a moment about the extraordinary complexity of the human eye, with its adjustable lens and all the various muscles involved in focusing and light control, as well as the complex retina and its nerves. All that has to be combined with an ability in the brain to interpret all the electrical data being sent to it, so that we can make sense of what we see!

What about the amazing nests of the weaver bird, which constructs a nest that hangs by a woven thread from a branch in a tree? The instinctive ability to construct the part of the nest in which the eggs are laid had to be there at the same time as the ability to construct the hanging rope that keeps it suspended. Neither part was any use to the bird without the other. It wouldn't help the bird to be able to construct the hanging thread until it knew how to attach the nest at the bottom.

A hanging weaver bird nest

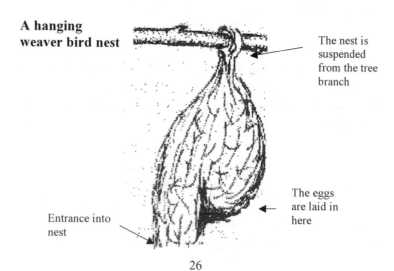

The nest is suspended from the tree branch

The eggs are laid in here

Entrance into nest

Remember, all this ability had to be programmed in some way into the brain of the bird, as the young birds breeding the following year would not have seen their parents building the nest in which they were born!

Another well-known and well-documented example is the bombardier beetle, which frightens away its enemies with an explosive defence mechanism! The biological systems necessary for the process involve the ability to manufacture two chemicals, hydrogen peroxide and hydroquinone, that explode when mixed together in the correct proportions. Because the reaction is so slow, however, it does not take place until an enzyme catalyst is added, which the beetle produces when it needs the reaction to occur. In addition, of course, in the nervous system of the beetle there must be control mechanisms, so that everything happens at the right time and it doesn't waste its energy 'firing off' when no enemy is threatening it! Everything has to be in place from the start, or the beetle gets no advantage from the system. There are hundreds of examples of such complex systems throughout the animal world.

In scientific circles there is a vigorous ongoing debate about intelligent design theory, and not all evolutionists are convinced by the scientific arguments advanced. Indeed, some oppose it precisely because it is being used, or can be used, by creationists to support a belief in a Creator God. An even more revealing objection to intelligent design by some evolutionists is the claim that it cannot be considered a scientific theory at all, since it contradicts a naturalistic understanding of the universe. In other words, the possibility of intelligent design pointing to intelligence operating in natural processes, or being involved in their beginning, cannot be true, since naturalism rules that out from the start!

This, of course, is not objective science, but is based on a strong religious conviction!

The greatest challenge facing the theory is to find some objective standard by which we can recognise what is designed and what has happened naturally. If we were to find a stone on a beach carved in the shape of a dog, we would probably conclude that some intelligence was involved in its production. But what if it is a perfectly rounded pebble, which may have been formed by the sea rolling that pebble against others until all its rough edges had been smoothed off? How can we decide whether intelligence is involved in the design or not? Can we formulate objective measures, by which we can differentiate chance and natural effects from designed or deliberate intervention? Some scientists have claimed to have designed mathematical means of calculating the probability of something happening by chance or by design. Others dispute this. No doubt the argument will continue for many years to come, and it is not yet clear whether intelligent design will win widespread acceptance or not.

Care please!
We should, however, note once again at this point that intelligent design theory cannot 'prove' the existence of the God of the Bible as the Creator of all things. We must never forget that it is by faith that we understand that the heavens were made. Hebrews 11:3 tells us: 'By faith we understand that the universe was formed at God's command, so that what is seen was not made out of what was visible' (NIV). Furthermore, God has told us how he did so in Genesis. Evidence of design, therefore, is exactly what we might expect to find on the basis of the biblical teaching of creation. The theory of intelligent design can be

welcomed as yet another indication that our initial assumption and belief that God is the Maker of all things fits in with the universe we live in and can study. Our 'faith' is reasonable, and the observations we make by looking at the natural world are consistent with the Bible's teaching.

An objection put forward against intelligent design as a theory is the existence of 'bad things' in nature. If God designed everything, then why do things like cancer and other illnesses occur? Where did parasites come from? Why are some creatures poisonous or dangerous?

This is another part of the great question regarding the origin of suffering in the world. If God is so good and is all-powerful, then why isn't creation perfect? The answer lies in the Christian doctrine of the Fall of Man and its consequences. This world has been cursed by God and has been altered from its original creation. Suffering and disorder are a part of the judgement and punishment given by God for Adam and Eve's disobedience and fall into sin. While we can still see the evidence of the hand of God in designing the world and the universe, we can also see the result of God's intervention to ensure that the earth remains a paradise no more. That it will not always be as it is will be followed up in the final chapter.

4
Genesis and genes

MODERN genetics is usually seen to have its beginnings in the work on inheritance in peas by the Austrian monk Gregor Mendel in the nineteenth century, about the time when Darwin wrote his *On the Origin of Species*, but it lay largely unnoticed until the beginning of the twentieth century. In Darwin's time the science of genetics, therefore, was still in its infancy, and little was known about how characters were passed on from one generation to another. Research since then has not only described the physical process of ordinary cell division in the development and growth of an organism, but also how genetic information is passed on from generation to generation through sexual reproduction.

It has been known for many years now that the genes, which are part of the chromosomes of plants and animals, contain a coded language strung together along the DNA double helix, which we now know as DNA or deoxyribonucleic acid. Comprised of only four different nucleotides, this basic language is amazingly efficient in passing on information to regulate the activities taking place within cells, but it also acts as a blueprint, like a set of plans for a building, of how the cells, and the larger organisms made up of those cells, should look and behave. Biologists have been able to unravel this DNA language, and read what bits of the chromosome direct and affect the different functions of the plant or animal in which they are found.

Normally each gene can exist in a number of different forms or alleles. These can be reshuffled or recombined to produce slightly different offspring in the next generation. Recombination of the genes makes it possible for there to be limited variation within the populations of each creature. Although changes in external shape and colour can be produced, the end product is basically the same as what you start with. An example is the domestic dog mentioned already in chapter 1. What tremendous variation is to be seen in the different breeds today, ranging from the St Bernard to the Mexican Chihuahua, or the Alsatian to the Poodle! All have been selectively bred from a couple of wild species of dog over the last 3,000 years or so. But despite the variation in size, shape, colour, etc., they remain still within the created dog-kind that God made.

Genes can only be recombined to a certain extent. Furthermore, recombination of the genes and natural selection working on them do not produce any new creatures or plants. Natural selection is not a creative influence on organisms. It can only act on and mould what is already there. The differences existing between Darwin's finches, which he described following his visit to the Galapagos Islands in 1837, can be explained on the basic of recombination and selection on the different islands with their different habitats. However, they remain still finches, most of whom are still capable of interbreeding. From a creationist point of view, all these finches have descended from one or two finch species belonging to an original finch-kind, which arrived in the past from South or Central America.

Mutations the great hope
Way back in the 1930s, it was discovered that ionising radiations like X-rays and some chemicals could cause new

changes in the chemical structure of genes, and these new 'mutated genes' or mutations then altered the way they acted in the affected plant or animal. At the time it was thought that this relatively simple process, of randomly ('by chance') produced variation in genes, would lead to more and more new mutations being formed. The beneficial mutations formed would then give an advantage to the organism containing them, and they would be passed on to their offspring, plant or animal.

And so the process would go on. Natural selection would automatically choose out the good mutations and weed out the bad or harmful ones, and in this way new evolutionary advances would be made. This was seen as something that happened over long periods of time. Eventually small evolutionary changes would add up to produce big changes with time.

This initial optimism was, however, to be disappointed. Studies on mutations have shown that the vast majority of them, instead of conferring some advantage on the organism containing them, are generally harmful or deleterious to its survival. Soon after the turn of the last century, geneticists began breeding the fruit fly, *Drosophila melanogaster*, and since 1910, when the first mutation was reported, some 3,000 mutations have been identified. All of the mutations are harmful or, at best, harmless; none of them produces a more successful fruit fly.

There are now flies with no eyes; flies with hairy backs; flies with smooth backs; flies missing parts of their wings; flies with extra parts like more legs or wings; flies with different-coloured eyes or none, and a host of other different-looking flies. None of the mutations, however, has produced a more successful fruit fly or anything other than a fruit fly. Although some of the changes mentioned seem relatively neutral, their effects seldom contribute to an enhanced survival of the affected individuals. This is what we would expect if God created the fruit flies with their variability in the beginning. These mutations are like a child, with a coloured pencil, making random changes to an original good blueprint of an aeroplane or a car. They are very unlikely to produce a more effective plane or vehicle.

Helpful mutations?
Very occasionally mutations do seem to be helpful to an organism. But these cases nearly always involve the loss of something, rather than anything new being made.

One example is the wingless beetles found on some islands. For a beetle living on a windy island, wings can be a definite disadvantage, because airborne beetles are more likely to be blown into the sea and drown. A mutation causing the loss of flight could be helpful in such a case, enabling the beetle to survive. At the same time, in other situations it could be harmful, exposing the beetle to attack from predators by depriving it of its normal means of escape!

Another example is that on the Galapagos Islands there are wingless birds, which have become good divers in order to survive. But other flying birds continue to survive on the same island, so being wingless isn't essential for survival after all!

Then there are sightless fish living in caves. Eyes are quite vulnerable to injury, and a fish that lives in the pitch darkness of a cave, having been carried in by a flood perhaps, might benefit from a mutation that caused the loss of its eyes, thus reducing the chance of its injuring itself. For a fish in its normal habitat in a world of light, having no eyes would be a terrible handicap, but it is no disadvantage in a dark cave.

While such mutations do produce a drastic and sometimes beneficial change, it is important to notice that they always involve the loss of genetic information and not the creation of new information. No mutation has been observed to cause the production of wings, or any other new organ where none normally exists.

It has been suggested that some resistance to some antibiotics, which has developed in bacteria, is an example of beneficial mutations appearing in those bacteria and thus enabling them to survive. However, investigations into the genetic changes that lead to antibiotic resistance do not support the common assumption that this is an example of evolutionary advance. These mutations consistently reduce or eliminate the function of normal chemical processes occurring in the bacteria. While such mutations can be regarded as 'beneficial', in that they increase the survival rate of bacteria in the presence of the antibiotic, they involve processes that cannot explain how these bacteria evolved in the first place. A high cost is involved for the bacteria in the loss of pre-existing cellular systems or functions, and such loss of cellular activity cannot legitimately be offered as a genetic means of demonstrating evolution.

The other thing to note about mutations, or indeed recombination of already existing variability, is that they are extremely limited in their effects. There is a limit in any change, and beyond this organisms cannot change. Plant breeders have, for example, been able to raise the percentage of sugar contained in sugar beet from around 6 per cent in the wild species to 17 per cent in the cultivated varieties. But that seems to be the limit beyond which no more improvement can occur. All the possible variability that God created in the sugar-beet kind has been gathered into one variety, and that is as far as we can go.

Comparisons of DNA

In the chapter on comparative morphology we looked at the similarity of different plants and animals. We saw that, rather than pointing to a common descent from earlier evolutionary ancestors, these similarities can be interpreted as pointing to a common designer. It should not surprise us that when we compare the DNA of different animals there are often striking similarities. Some studies have shown that human beings share up to 97 per cent of their DNA with apes, though other researchers now suggest that the similarity may be slightly less than 95 per cent. Does this prove that humans have 'evolved' from a common ancestor with chimps? Not at all! We should not forget that if only 5 per cent of the DNA is different, this still amounts to 150,000,000 DNA nucleotide base pairs that are different between them. Think of them like letters of the alphabet. They are equivalent to approximately 12 million words, or 40 large books of information!

Summing up

The important question we have to ask is whether the genetic mechanisms that can give rise to new varieties and, sometimes, if they can't breed with existing ones, new species of plants and animals, can explain how the original families and genera arose. In other words, are the limited changes that we see taking place today, and can explain on the basis of modern genetics, sufficient to explain the General Theory of Evolution? The answer is a resounding no! Where does the process of speciation, the making of new species, lead?—to the eventual death and extinction of plants and animals. They change and adapt to new habitats through recombination of their genes, acted on by natural selection, until they can adapt no further. They reach the genetic limit beyond which no more change is possible. Mutation doesn't help, as its effect is limited and normally harmful, not helpful, to the organism.

5
Fossils: what are they saying?

Fossils are the remains of plants and animals that have been turned into stone, or the imprint of such organisms in stone. Strictly speaking, of course, coal and oil are also examples of fossils that have been chemically changed, and they are therefore known as fossil fuels. Millions of fossils have been found all over the world. Many of them are of creatures and plants now extinct. Others are of recognisable species still living today. More species are identified every year.

What is quite remarkable is the number of living fossils: plants and animals living on the earth today which were thought to have become extinct a long time ago. An example is the coelacanth, a fossil fish that was supposed to have died out about seventy million or more years ago. That status had to change after a living coelacanth was discovered in 1938, and many others have been caught since. Their last appearance in the fossil record may lie millions of years ago, according to evolutionary teaching, but here they are still surviving today. Why didn't they leave any trace in the rock strata formed over all those millions of years, if they didn't die out as first thought?

Those troublesome gaps
Now what does a study of fossils (palaeontology) tell us about the past history of this planet? If evolution was true and if, for example, all animals were related by descent,

then, assuming that the fossils and the layers in which they are found were formed over millions of years, the fossil record would be expected to show a series of animals slowly changing from one 'form' into another. Fossils should 'document' the slow but sure advances and changes in organisms over time. There should be evidence of creatures changing from fish into amphibians, reptiles into mammals, and so on.

It was recognised early on that this is not what the fossils seem to point to, since there are large gaps in any such postulated series of fossils. A few fossils have been put forward as transitional fossils, such as the famous Archaeopteryx, suggested to be midway between reptiles and birds because it was claimed to have some 'reptilian' and some 'avian' features. Today, however, even that is usually considered to be just a fossil bird, and its 'reptilian' features are found in either modern or extinct bird species. In most of the theoretical trees of evolutionary descent, there are huge gaps where no fossils are to be found.

In the second half of the nineteenth century and through most of the twentieth century, it was anticipated that further discoveries would be made that would fill in the gaps or missing links in the fossil record. As one scientist wrote back in 1955:

> That evolution did occur can only be scientifically established by the discovery of the fossilized remains of representative samples of those intermediate types which have been postulated on the basis of the indirect evidence. In other words, the really crucial evidence for evolution must be provided by the palaeontologists, whose business it is to study the evidence of the fossil record.[1]

---

1. Le Gros Clark, *Discovery*, Jamuary 1955, p.7.

As Huxley, an enthusiastic proponent of evolution, wrote in the nineteenth century: 'If evolution had occurred, there would be its proof: if evolution had not occurred, there would be its refutation.'

Such was the view still when I was in university in the 1960s.

The gaps are real

That hope has been disappointed, and all but abandoned, by many researchers today. In discussing the origin of the fishes, Ommaney wrote:

> How the earliest chordate stock evolved, what steps of development it went through to give rise to truly fish-like creatures, we do not know. Between the Cambrian, when it probably originated, and the Ordovician, when the first fossils of animals with really fish-like characteristics appeared, there is a gap of perhaps 100 million years which we will probably never be able to fill.[2]

When the fossil record is examined, practically all orders or families known appear suddenly, and without any apparent transitions from other life forms. These gaps in the fossil record showing how new species developed have been remarkably difficult to close, so much so that some scientists have come to the conclusion that they are part of the fossil record itself and not a result of poor sampling or bad luck on behalf of those seeking to find the missing links.

---

2. Ommaney, F. D., *The Fishes* (New York: Time Life, 1964), p 60.

Punctuated Equilibria

The acceptance that these gaps are real has become so compelling, even among some evolutionists, that the last decades of the twentieth century have seen the emergence of new theories as to how life evolved.

The *Theory of Punctuated Equilibria*, for example, suggests that the process giving rise to new species was not slow and gradual, as previously thought, but rapid and restricted to small areas of the earth's surface. For long periods of time, life continued in a stable equilibrium; but then, suddenly, this equilibrium was punctuated by periods of rapid evolutionary advance. It is suggested that in certain places, and over relatively short periods of time, conditions were such that some animals, for example, changed quickly from one form to another, leaving behind little or no trace of the change in the fossil record.

Punctuated Equilibrium Theory of Evolution

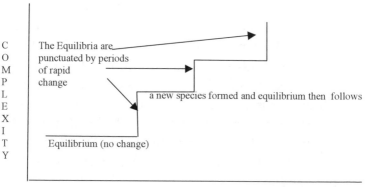

This theory postulates that the evolutionary development from one species to another took place in localised regions of the earth; was very rapid with big changes in a short time, thus leaving behind no fossils to discover. Most of the time little or no real change takes place (Equilibrium continues).

In other words, evolution is regarded no more as a gradual ascent up the evolutionary slope, but as a climb up a staircase with steps of different height and at different spacing. The changes were so rapid and so localised that the possibility of fossils being formed was almost infinitesimal. Hence their absence in the fossil record.

If the question is put as to how these changes happened, it is admitted that we have no idea at present. Certainly the science of genetics gives no known mechanism for bringing it about. Nevertheless, we are told that this is how it must have occurred, and that 'no doubt' further research will uncover the mechanism. If evidence is requested in support of this theory, then it is the absence of transitional fossils that is put forward as the proof!

Let me emphasise that the reason such theories have arisen is primarily due to the lack of expected transitional fossils—these gaps in the fossil record. It is because many now expect that they will never be found that such new theories are put forward, to allow for 'evolution' in its classic sense still to be considered possible.

But let us look at the fossil with different presuppositions. Let us assume that the Bible is true, and that God made the different kinds of plants and animals quite distinct from each other from the beginning. What would be our expectation in looking at the fossil record in regard to transitional fossils? We would not expect to find any, of course, since they never existed. Now that doesn't *prove* absolutely and beyond any doubt that our initial assumption is true—as the existence of alternative theories shows —but it is at least what we would expect on the basis of what the Bible teaches. Fossils do not contradict the teaching of the Bible.

A few Fossil puzzles

As well as the well-known gaps in the fossil record, palaeontology gives us other reasons for doubting the general theory of evolution. One of these arises from the study of palynology. Palynology studies fossil remains through a microscope. Palynologists have found parts of plants, too small to be seen by the naked eye, in rocks supposed by evolutionists to be too old to contain them. Fossil pollen grains have been found in Precambrian rocks, for example, far too old on evolutionary dating to contain them, since plants were not supposed to have evolved at that stage! This would not be a problem if life didn't evolve but was created by God in its vast variety from the beginning.

A further problem for evolutionary theorists is the occurrence of upside-down strata, found all over the world and in every continent. These are layers of rock containing fossils that occur in the wrong order. The expectation, if evolution is true, is that the deepest or lowest rocks in a succession of rock strata would contain the oldest fossil forms in the development of life, and that the highest strata would contain the newest or youngest forms. Now it is true that, generally speaking, lowest layers of rock often contain fish and other creatures that live in water, and the highest strata contain more land animals and mammals, although sometimes they are all mixed up together. Reasons why that should be so are suggested in the next chapter, when we shall look at how fossils were formed.

Returning to upside-down strata, one such example is the Matterhorn in Switzerland. In order to explain why its rock layers are in the wrong order to what is expected, it is postulated that the entire mountain was thrust up over other younger rocks from some 30 to 60 miles away (an over-thrust)! Travelling overland for those long distances, it eventually arrived without leaving any evidence of the grinding pathway it ought to have left in its wake! And the Matterhorn is only one of a number of Swiss mountains that are out of the standard geological order. They all had to be moved into position from huge distances away. Another massive mountain in the Swiss Alps is the Mythen Peak. It is claimed that the Mythen moved all the way from Africa into Switzerland! In this mountain the Eocene strata (supposedly 55 million years old) are lying under Triassic rocks (225 million), Jurassic (180 million), and Cretaceous (130 million) strata. According to the theory of evolution, the Eocene is supposed to be on top of the Cretaceous, Jurassic, and Triassic—but instead it is under all three!

Of course, evolutionist geologists have theories—some seeming very far-fetched indeed—to explain away these anomalies. But if the geological column is regarded not as a record of evolutionary progress over millions of years or as a proof of evolution in progress, but as the result of the sorting and destructive actions of the worldwide Flood of Noah, these explanations are not necessary at all. Even if some degree of overthrusting has occurred, as some creationists (but not all) accept, the extent of it is such that it is impossible to explain it on the basis of present-day geological processes. Something extraordinary like the Flood of Noah must have been involved.

Human artefacts in rock

Another puzzle sometimes put forward by creationists is the possible discovery of human remains in rock strata reputedly too old to contain them. These remains range from human footprints, found in many locations all over the world in ancient sedimentary rocks, to human bones and artefacts in rocks far too old to contain them.

43

What could be the oldest fossil footprint ever discovered was found in 1968 by William Meister, an amateur fossil collector. The print appears to be the impression of a sandalled shoe crushing a trilobite. According to the theory of evolution, the trilobites died out some 300–600 million years ago! Meister made his find during a rock and fossil hunting trip to Antelope Spring, Utah. He had already discovered several fossil trilobites when he split open a rock with his hammer and made the find. The rock fell open, he said,

> like a book, revealing on one side the footprint of a human with trilobites right in the footprint itself. The other half of the slab of rock showed an almost perfect mould of the footprint and fossils. Amazingly, the human was wearing a sandal.

Evolutionists suggest that the fossil must have been formed through natural processes, though what kind of process they are unable to say.

A more famous case is the study of fossil footprints in the Paluxy river bed in Texas, about which a creationist film was made some years ago. Here, however, a note of caution is in order. Subsequent research has indicated that these may not be human footprints after all. Some creationists who championed these as human footprints have changed their opinion. This illustrates something very important. Sometimes creationists (and of course evolutionists) may come to a conclusion about some evidence which later they have to revise in light of further discoveries. That does not mean that their presuppositions are wrong, but their interpretation of the evidence was. Even if the Paluxy river bed footprints turn out not to have been made by humans after all, it does not prove that evolution is true and creation is false. It does mean, however, that creationists cannot and should not use such evidence as evidence for the correctness of their theory of origins.

There have been numerous reports of other human artefacts found in rock strata too old to contain them, such as gold chains, hammers, even human bones. However, unless it can be conclusively shown that they did originate in the older strata where they are reputed to come from, their use as an indicator of creationism is very limited or useless. A gold chain that has been released from the rock that supposedly encased it is no evidence at all for a recent creation, since it cannot be checked by others. It is important that those who seek to point to evidence contradicting evolutionary theory are ruthlessly honest and careful in their own evidence-gathering and interpreting! Our God is a God of truth and does not need falsehoods to support his Word.

My own view is that an honest study of fossils does not support the theory of evolution but gives evidence of God creating the different kinds of plants and animals quite distinct from each other. Of course that is no surprise for a Christian. A person who believes in the God of the Bible and who is trusting God's Son, our Lord Jesus Christ, as his personal Saviour, should have no problem in believing that God speaks the truth in his Word!

6
Fossils:
a testimony to catastrophe

FORTY years ago the prevailing view about the rock strata of the world was that they had been laid down gradually over millions of years. We were told that studies of rock formation today suggest that it takes many years to form even a relatively thin sheet of rock, and therefore strata that are thousands of feet thick must have taken a very long time to form. The geological column claiming to show the different ages of the earth, and the fossils associated with those periods, is measured in hundreds of millions of years, going back, it is said, to the forming of the earth about four billion years ago.

What we were not told, however, is that that very column, showing the development of life through the fossil-bearing strata and geological ages of the earth, and assigning dates of hundreds of millions of years to its different parts, is itself an artificial theoretical construction, based on the assumption that evolution had occurred. Even if it is granted that the succession of fossil-bearing rocks around the world is roughly in accordance with the geological column—fish in the lowest strata, mammals in the highest, etc.—the slow, gradual process of their formation over millions of years has not been proved and is not the only possibility.

Ancient mountains?
The great question here is quite simply, How old are the rocks of the earth? How can they be measured? Clearly we

cannot rely on human observation, since no one has been around long enough to see them being formed—or so evolutionists tell us. According to evolutionary theory, human beings only appeared in the last two to three million years (depending on how you define a human being in the postulated ape-to-human sequence of development). If no one was there at the time to record the past history of the planet, it is surely logical to construct it on the basis of what we see happening today.

This sounds very plausible and sensible, except for one thing. We do have an eyewitness account of the past history of this planet in the Bible. The story of the universal Flood in the time of Noah is found in the Old and New Testaments of God's Word. Jesus spoke about it as a historical event (Luke 17:27) from which we should learn lessons about being ready for the future judgement of the world at the return of Jesus Christ to this world at the end of the age. The apostle Peter even warned us that the time would come when people would deny it happened (2 Peter 3:5-7) and live as if God had never judged this world in the past.

Just as the discovery of the Bible's teaching about the creation of the world shook my belief in the theory of evolution and caused me to reconsider the evidence for it in the light of that teaching, so the story of Noah's Ark and the worldwide, catastrophic Flood meant taking a new look again at the age of the earth. I freely admit that it is not an easy subject to look at. There are observations that seem to point to a great age for the earth, and we shall consider some more of them in the next chapter. The sheer depth of sediments in some parts of the world and the signs of deep erosion, such as in the Grand Canyon in America, seem to suggest that these changes in the planet's surface must have required a very long time to come about. By today's reckonings, mountains grow (or are eroded away) very slowly, so if it happened at today's speeds, how many years did it take to lift up the great mountain chains of the world?

But here in this question an assumption underlying all dating methods is once more clearly revealed. It is known as *uniformitarianism* (see pages 13-14): that is, the assumption that the rate at which things happened in the past can be assumed to be uniform, or the same as the rate at which things happen in the present. But is that a reasonable assumption? Were the mountains, for example, built in the past at the same rate as mountains are built today? Or are the present-day rates of erosion caused by the River Colorado in the Grand Canyon the same rates as have applied over millions and millions of years? The short answer is no!

The Genesis Flood

If the Bible is true, then the story of the worldwide catastrophe in the time of Noah has a profound implication for the biological and geological history of this planet. While the details of the effect of the Flood in modern geological terms are not given in any detail in the Bible, it is clear from the account in Genesis that the flood waters covered all the land masses of the world to a considerable depth— so deep that all air-breathing creatures, apart from those in the Ark, perished. If that is true, then we should expect to find water-borne sediments all over the globe, even where there is no evidence of water laying down any strata today. This expectation is confirmed by the occurrence of fossils and marine sediments even at the top of the highest mountain ranges in every continent of the world. At the same time, if the Flood did take place, then the assumption of uniform rates of rock strata deposition and erosion is simply not valid and true. The scale by which many geological events in the past happened is far, far greater than that observed today. It needs therefore to be remembered that any attempt to reconstruct the history of this planet in biological or geological terms, if it does not take into account God's testimony to its past history, is bound to lead to false conclusions.

Two mechanisms that caused the global flood are mentioned in the Bible. One was the breaking up of the 'fountains of the deep', and the other the falling to earth of rainfall for forty days and forty nights. Various theories have been suggested by creationists as to what these phrases mean in geological and physical terms, but all are united in their portrayal of catastrophic effects upon the physical surface of the globe.

I was first influenced in my own thinking over the geological history of the earth by reading, in 1964, Morris and Whitcomb's ground-breaking book *The Genesis Flood*. Creationist and evolutionary theories have developed a great deal since this book first appeared in 1961, so it is not surprising that it is now somewhat dated and in need of a revision as far as the scientific theories are concerned. Their book, however, demonstrated that the various attempts made to redefine the biblical account of a global flood as some sort of local catastrophe do so at the expense of sound hermeneutics and biblical orthodoxy. The book also presented a working and intelligent model of biblical catastrophism that showed how a global flood provided a biblical, sound and reasonable answer to the question of the age and the geology of the earth.

However, any suggested model for the Flood that is put forward by creationists today can only be speculative and not final. We cannot do an experiment to repeat it, and the detail given in the Bible is insufficient to be sure of how it happened. As I have emphasised a number of times, you cannot prove these things beyond all doubt in a laboratory. But as, by faith, we accept the Bible's teaching, we discover that there are good grounds to believe them. One piece of evidence for the Flood is the rapid formation of fossils.

Rapid fossilisation
One of the interesting things about the fossil record is the evidence of creatures that died quickly and immediately before being fossilised. That is, of course, partly what we would expect, whether we accept evolution or creation, since normally, unless an animal is quickly covered over by sediments, it rots away and disappears. Generally speaking, when we find a fossil of an isolated tooth or shell, for example, it is not possible to say how quickly or slowly it formed. However, there are countless examples of fossils concerning which it is obvious that long time-spans could not have been involved in their formation.

There are examples of fish dying in the act of swallowing their prey or being buried in their death throes. By the way, since marine creatures would be the most likely ones buried by the sediments carried by a worldwide flood sweeping over the earth's surface into the early oceans—land creatures would be more likely to disintegrate and rot as they floated in the waters—we would expect most fossils to be marine fossils. And marine fossils do indeed represent more than 95 per cent of the fossil record. Fish-bearing rocks would also tend to be the lowest strata, since they would have been buried first.

However, even on land, fossils show many animals which died rapidly and without warning. For instance, there is an exquisitely preserved fossil of an extinct marine reptile called an ichthyosaur. The mother ichthyosaur is fossilised, having almost completed giving birth to a live infant—the beak of the young reptile is still inside the mother's birth canal.

Many other fossils have soft tissue features so beautifully preserved that they must have been buried and hardened before they could be damaged by scavengers or decay. Well-preserved fossils of jellyfish have been found. Jellyfish mainly float in the water and are therefore at the mercy of predators when they die. If you find a jellyfish on a sandy beach, you will notice that it rapidly loses its structural details as it decays. Yet in the fossil record there are many examples of perfectly preserved jellyfish, showing that they must have been rapidly buried and fossilised. In such cases the normal evolutionary view, that fossils are formed by sediments slowly covering up dead animals, is completely inapplicable.

In many places a number of different types of fossil—marine creatures and land animals and insects—are all found together in large 'fossil graveyards'. It is sometimes suggested by evolutionists that they must have somehow lived together in an area where land and sea were closely associated. However, the fact that they are found together in one place may mean nothing more than that they were buried together—something that can happen in the event of a massive flood sweeping them together into heaps before covering them with sediment. This sort of mixing explains how creatures from widely separated regions may be carried along by water before being dumped where they now are. Many such accumulations of fossils are overwhelming evidence for rapid burial and hardening in some great catastrophe.

How old?

The standard view of fossils is that they are very old. Yet again there are many examples of fossils that appear much younger than their anticipated age. In some cases unfossilised wood, or partly fossilised wood, has been found in coal seams that are reputedly millions of years old. Dinosaurs have been supposedly extinct for over seventy million years, yet some have been found with intact, unfossilised blood traces and DNA.

It is claimed that stalactites in caves have taken thousands or even hundreds of thousands of years to form. Such ages are derived from measuring their growth rates today. But we have no way of knowing how fast they formed in the past, particularly in the times of greater rainfall that would have followed the Flood. As rainfall decreased, so would water inflow into the cave systems, and hence growth would have slowed.

Stalactites do not necessarily need a long time to form, as is seen from those hanging under motorway bridges and in some of our modern concrete tower blocks! All they need is sufficient acidic water flowing through carbonate-containing rock layers to dissolve out the carbonates, which then crystallise out in the caverns as the water evaporates in the air. In one study this was replicated in a laboratory after two days.

Other studies have shown situations where a stalactite has grown about twenty-five centimetres in a period of ten years. Another stalactite has a preserved bat within the formation, and cases of hats and other articles being 'fossilised' into stone in a period of a few years are also documented. In other words, estimating the age of a stalactite or a stalagmite from the rate at which it grows today is no proof that it is that old.

In the next chapter we look at radioactive dating and other methods that seem to point to a very young earth and universe.

7

What about radioactive dating?

INEVITABLY, whenever the age of the earth or universe is mentioned, someone asks about radioactive dating. We often hear about a fossil that is 'dated' as millions of years old, plus or minus a few million! How reliable are these dates? If only one of them were correct, then the whole belief in a young earth of only thousands of years collapses immediately. There are also other observations that seem to suggest that the universe is far older than a few thousand years, and these we will consider later.

Radioactive dating methods can be divided into two main groups: those that deal with relatively recent dates and those that are concerned with dates in their millions. The most well-known short-term method is radiocarbon dating and one of the best-known long-age methods is potassium-argon dating. All methods, however, rely on the same basic principle. Radioactive elements are unstable and break down or decay into different chemical elements. The amount of different elements can be measured very accurately and the ratio of different elements calculated. Knowing the rate of the decay process, we can then calculate how long it has taken to arrive at these ratios.

Radiocarbon dating

First, let us look, for example, at radiocarbon dating. Nitrogen atoms in the upper atmosphere of the earth are hit by cosmic radiation and create C^{14} (carbon 14) atoms. They are quite rare atoms: one radioactive C^{14} atom exists for every one trillion stable C^{12} (carbon 12) atoms. These atoms rapidly react with oxygen in the air, forming radiocarbon-dioxide which, along with ordinary carbon dioxide, is absorbed by living plants and becomes part of their living tissues. Gradually the radiocarbon breaks down into ordinary carbon at a constant rate, so the C^{14} is constantly disappearing and being replaced with ordinary (C^{12}). Ordinary carbon is stable and does not decay any more. Calculations have shown that half of any quantity of radioactive carbon will disappear in 5,568 years. By measuring the relative amounts of C^{14} and C^{12} in a sample of plant tissue (wood, for example), it can be calculated how long it is since the plant died and stopped assimilating carbon dioxide from the air.

For radiocarbon dating to be reliable, scientists need to make a number of vital assumptions.

Firstly, it is assumed that C^{14} decays at a constant rate. However, some experimental evidence indicates that C^{14} decay has slowed down from the rate in the past and that, millennia ago, it decayed much faster than is observed today. Any suggested dates of organic matter older than a few millennia would therefore appear to be older than they really are.

Secondly, the theory behind C^{14} dating demands that there is the same rate of cosmic production of radioactive carbon throughout time. This is an unproved assumption and is unlikely to be correct if the magnetic field of the earth was much stronger in the past than it is today. There is evidence to show that the strength of the earth's magnetic field—which would deflect cosmic rays away from the earth—was indeed much greater in the past compared to recent times. That would lead to fewer C^{14} atoms being formed previously and would make carbon dates in archaeology, particularly past a couple of thousand years, seem to be older than they really are.

Thirdly, the environment in which the artefact being measured lies has a great effect on the apparent rate of decay. For example, C^{14} leaches at an accelerated rate from organic material which is saturated in water, especially saline water. Any waterlogged tissue will give an older date than it really possesses.

Fourthly, for C^{14} to give an accurate result, the thing being measured must have been protected from contamination. Organic matter, however, being porous, can easily be contaminated by organic carbon in groundwater. This increases the C^{12} content and interferes with the ratio of C^{14} to C^{12}, once more giving an older age for the measured material than it really possesses.

I have spent some time examining radiocarbon dating because the problems in the method and the underlying assumptions are equally valid for other methods of radioactive dating. That they are valid objections is seen in some bizarre results obtained by the radiocarbon dating method.

Examples of abnormal C^{14} results include testing of recently harvested, live mollusc shells from the Hawaiian coast showing they had died 2,000 years ago, and snail shells just killed in Nevada, USA, dated at 27,000 years old—a long way past their sell-by date! A freshly killed seal at McMurdo Sound, Antarctica, yielded a death age of 1,300 years ago, while radiocarbon dating of a petrified miner's hat and wooden fence posts, unearthed from an abandoned nineteenth-century gold-hunter's town in Australia's outback, showed that they were 6,000 years old.

Recently, radioactive carbon has been found in 'billion'-year-old diamonds, dinosaur bones and many, if not most, biological fossils that still contain any carbon at all, even though they are supposed to be millions of years old.

Potassium-argon dating

Other methods, such as potassium-argon (K-Ar) dating, which is used to date 'old rocks', are equally problematical. There have been a number of cases where rocks of known age—that is, formed within historical time, mainly of volcanic origin—nevertheless give ages of millions of years with this method. One example is from solidified lava at Mount St Helens volcano in America, which exploded in 1980. Although we know the rock was formed then, the rock was 'dated' by the potassium-argon (K-Ar) method as 350 thousand years old! Another example of K-Ar 'dating' of five lava flows from Mt Ngauruhoe in New Zealand gave 'dates' ranging from less than 0.27 up to 3.5 million years. But one lava flow occurred in 1949, three in 1954, and one in 1975!

What seems to have happened in this case was that some radiogenic argon (Ar^{40}, formed from the decay of the radioactive potassium) from the molten rock spewed out by the volcano was retained in the rock when it solidified. The scientific literature lists many examples of excess Ar^{40} causing 'dates' of millions of years in rocks of known historical age. This excess appears to have come from the upper mantle of the earth, below the earth's crust. This is consistent with a young world—the argon has had too little time to escape into the atmosphere since the creation of the world. Now if extra Ar^{40} can cause exaggerated dates for rocks of known age, then why should we trust the method for rocks of unknown age?

Another problem that often arises is the different dates arrived at by using different methods. Obviously if two methods give two different dates, then at least one of them must be wrong. In Australia, for example, some wood was buried by a flow of lava and was consequently charred by the heat. The wood was 'dated' by radiocarbon (C^{14}) analysis at about 45,000 years old, but the basalt was 'dated' by the K-Ar method at about 45 million years old! Some fossil wood from Upper Permian rock layers, supposedly 250 million years old, has been found with C^{14} still present, even though the C^{14} would have all disappeared if the wood were really older than 50,000 years, let alone that great age.

An interesting question arises here, however. How is it that in scientific literature we often see ages given which confirm the evolutionary date already assigned to a particular fossil or rock stratum? It is not due to inaccurate measuring of radioactivity in the laboratory! It arises either from deficiencies in assumptions used in the method of calculating the age, as above, or sometimes from a more subtle form of bias.

How old did you say?
Often a sample of rock will yield a number of results by radioactive dating, and these are considered. Let us suppose, for example, that a rock sample yields an age of two million years for a rock that, according to evolutionary dating, is supposed to be 200 million years old. Our efficient lab scientist gives his results to the field geologist who sent the sample in. The geologist knows roughly what he expected to find—let us suppose it should be about 200 million years old. Two million years is then far too young! He speaks to the lab worker, and it is 'obvious' then that the sample must somehow have been compromised, either by the loss of some elements through leaching or diffusion out

of the sample, or because other elements have contaminated it in a similar way by coming into the sample. Now this is not deliberate dishonesty, but the normal bias every scientist has in dealing with his work. He expects to find what he is told to expect! It is only after repeatedly getting such false results that other possibilities are forced into his thinking. (We have seen this with the rise of punctuated equilibrium thinking in chapter 4. The old mantra of 'the missing fossils will be found one day' was repeated until the pressure of evidence compelled a rethink and new theories were proposed.)

Eventually, the laboratory result will yield a result compatible with that expected, since all the unexpected ones will be rejected as false! It is a self-confirming cycle. Whatever doesn't fit must be wrong, so only results that do fit are allowed.

Even the margins for error can be misleading. A date of 200 million, plus or minus five million, years sounds quite an accurate measurement of age! But it reflects simply the variation in the actual measurements of radioactive and other elements in the sample being tested and not an actual variation in age at all. It seems to add accuracy to the estimate of age but is in reality just an illusion.

Some Young Age measurements
In a brief book like this, we cannot look at all the various methods of measuring age that suggest that the earth is only young and not old. One researcher listed over fifty different methods of measuring the age of the earth that yield young ages. In mentioning these, however, it is worth pointing out that these methods, like all dating methods, rely on certain unprovable assumptions and therefore none of them 'prove' how old the earth really is!

I mentioned above the evidence that suggests that the magnetic strength of the earth has been declining for centuries, and possibly by as much as 40 per cent in the last thousand years. Calculations show that, if true, then the earth cannot be older than about 10,000 years, or it would have melted under the strong fields operating then. Of course, evolutionists have objections to such claims, but their counter-claims are based on unprovable assumptions too. Suffice it to say that the magnetic field decay is one good indicator of a young earth.

Another pointer to a recent creation is the rate at which the moon is receding from the earth. It has been known for some time that the moon is slowly drawing away from the earth, at about forty centimetres every decade. This rate would have been greater in the past. It doesn't take a genius to calculate that at this rate, even if the moon started in contact with the earth (an impossibility), then it would need less than two billion years to reach its present distance. This is the greatest age of the moon—not the age that it really is. It is also far too young for evolutionary theories of the origin of the solar system and is even much younger than the dates given to moon rocks by radioactive dating methods.

Helium is a very light gas formed by the decay of radioactive elements in the earth's crust, and it is constantly being produced. Some of it diffuses out of the atmosphere into outer space. However, if the earth really is billions of years old, there is far too little helium in the atmosphere. In fact there is only 0.05 per cent of what we would expect if the earth really is very old. Most of the helium that has been produced is still trapped in the rocks of the earth and should have diffused out by now, if the rocks really are millions of years old. Of course, we do not know how much

God created originally in the earth, so a calculation of the age of the earth is still not really feasible using this method.

Every day salt is pouring into the sea through the rivers of the world. Although some disappears again through precipitation and wind spray, it is still accumulating much faster in the oceans than it is escaping from them. The sea is getting slowly saltier! But here is the problem for those who believe in an ancient earth. The sea is not nearly salty enough for this to have been happening for billions of years. Assuming the most extreme limits in favour of an old ocean, the oceans cannot be more than sixty-two million years old, and that is much younger than the billions of years postulated by evolutionary theory. This does not, of course, mean that the sea is as old as that. God may well have created the original water sources of the original pre-Flood world with some salt in them originally. However, their present-day content is more consistent with biblical assumptions than evolutionary!

Finally, let me mention the case of supernova. A supernova is the result of the explosion of a massive star. One can shine so brightly that it briefly outshines the rest of the galaxy in which it is found. They are not rare to see in the sky. On average, our own Milky Way Galaxy should produce one supernova about every twenty-five years. After the explosion, the huge expanding cloud of debris is called a Supernova Remnant, which should be visible from earth through telescopes. A well-known example is the supernova in the Crab Nebula, which was so bright that it could be seen during daytime for a few weeks in 1054. Like any explosion, a supernova expands outwardly into space for hundreds of thousands of years, getting bigger and bigger. However, when we look into space, we can see evidence only for supernova that have grown relatively little. If the universe is billions of years old, we would expect lots of examples of huge supernova remnants all over the sky. There aren't any! Their absence points to a young universe and not one that is millions of years old.

In conclusion, in my opinion there is no conclusive proof that the universe is billions of years old. There are plenty of indicators of a young earth to support a belief in a recent creation. All dating methods rely on assumptions however and, depending on your assumptions, a case can be made for either!

8
Where do humans come from?

THE question about origins has occupied and interested people throughout history. The very fact that we are here compels us to ask where we come from.

The standard evolutionary answer is, of course, that we are here by chance and have gradually evolved from lower forms of life over millions of years. Our closest evolutionary relatives are supposedly the anthropoid apes, and a great amount of effort and expense has been expended in search of our evolutionary ancestors. A number of palaeontologists have pursued the search for the fossils of our predecessors with dedication and great zeal, and great reputations have been built on their discoveries and theories. Museums frequently show displays of our forebears living in prehistoric landscapes, and illustrations of the gradual changes from ape to man adorn school textbooks and popular science books. The question I want to ask in this chapter is, How true are these illustrations and displays? What about all those ape-men fossils?

First, we need to note how misleading words can be. Some fossils have been given the name Pithecanthropus— made up of Pithecus (ape) and Anthropus (human), thus meaning Ape-man. However, as soon as people name a fossil in this way, they have already accepted the reality of evolution! The very term 'ape-man' applied to a fossil presupposes that these creatures existed. In reality, all ape-men fossils are simply fossils that have been classified by evolutionists as transitional forms between apes and men.

Now, no one disputes that these fossil bones exist: the argument again is how we interpret them. To label them as ape-man fossils reveals an *a priori* commitment to evolutionary theory. If we call them simply extinct ape fossils or chimpanzee fossils or, in some cases, human fossils, then immediately their 'proof-value' disappears! What is interesting is to see how quickly a new fossil is claimed as a transitional form; yet, with time and more research, it is usually renamed and consigned to a clear biological type of creature, and its 'importance' declines.

Neanderthal man

For a long time Neanderthal man—so called because their first fossils were found in the Neanderthal valley in Germany—was heralded as a primitive type of human, just evolving from more apelike ancestors. They were claimed to be 'beetle-browed, barrel-chested, bow-legged brutes'. Artists' impressions showed pictures of them as typical stooping cavemen, hairy and with apelike expressions, usually dragging their female partner along by her hair. Most of it was pure speculation and imagination!

Further investigations and discoveries have revealed that they had slightly bigger brains than the average person today. They were intelligent and religious and could play musical instruments. They buried their dead with grave-goods and were excellent hunters and artists. Certainly they had some facial features that were different from most people today, but these can be largely accounted for through disease and environmental factors. Dressed in modern clothes and given a shave and a haircut, they would pass unnoticed down the average suburban street!

The fact that they tended to live mainly in caves is a reflection of the harsh climate in which they lived after the

Flood of Noah. Following that catastrophe, the world under-
went a period of climate change lasting for many centuries,
and those having to face the cold conditions that accompa-
nied the ice age that followed in the northern hemisphere
inevitably found it hard to survive. Neanderthal man repre-
sents a human grouping that has largely died out, although
biblical references to 'giants' suggest that for some time
genetic descendants of some of them may have existed in
parts of the Middle East up to fairly recent times.
Evolutionists today usually classify Neanderthals as *Homo
sapiens*—just a different racial grouping from most of us. In
a similar way Australian aborigines were once classed as
subhuman by earlier evolutionists.

Australopithecus

In looking at these so-called ancestors of humans, the number and variety of Australopithecus species identified so far
seems at first impressive. Many of these, however, depend
on a handful of bones found in small confined localities of
the earth. Time and time again, more sober assessment by
other scientists of the claims made by their enthusiastic
discoverers has led to reclassification of many of them
into either distinctly human or apelike species. That
chimpanzee-like and apelike species have existed in the
past and are now extinct, no one doubts. That they are human
ancestors is a very different thing, however. It provides no
evidence in support of the theory of evolution when that theory is presupposed to label them in the first place.

At present much attention is focused on the
Australopithecines (southern-apes) discovered by the
Leakey family and others. These fossils have been found in
Africa alongside those of other animals such as rhinocer-
oses, monkeys and hippopotamuses. One of their special

characteristics is that they were thought to have walked upright, even though their other features are apelike rather than human. Other scientists, however, have challenged the claim that they walked in a human way at all; they may have been knuckle-walkers like some species of ape today.

Actually it was not the skeletal features that attracted attention to the Leakey finds in the first place. It was the discovery of tools with the fossils, largely indistinguishable from modern human stone ones. Since the tools were found with Australopithecus, Louis Leakey assumed that that creature had made the tools. Some years later, however, Richard Leakey, Louis' son, found beneath the bones his father had unearthed 'bones virtually indistinguishable from those of modern man'. That probably solved the tool-maker mystery! If the bones of humans are found beneath those of their supposed ancestor, then something is wrong with the theory in the first place!

The Canadian Broadcasting Corporation produced a series of programmes in 1981 on Creation–Evolution. It opened with a medieval princess looking for something as she wandered in a castle garden. On a rock ledge by a pond there was a frog with big bulging eyes. The princess went to the frog, leant over and kissed it. Stars sparkled across the TV screen and suddenly a handsome prince appeared. As prince and princess embraced, the narrator stepped into the scene with this introduction: 'If you believe a frog turns into a prince instantly, that's a fairy tale; if you believe a frog turns into a prince in 300 million years, that's evolution.'

9
How did life start?

ONE of the greatest problems for those who believe in evolution is to explain how life started by chance in the first place. The more that has been discovered about the complexity of life at the cellular level, the more difficult it has become even to try to suggest a possible way for life to have spontaneously arisen on earth.

It is not just that there is an astonishing variety of complex organic molecules in cells, but they are interrelated by numerous different pathways and systems. Even a small change in some components means that the system cannot function at all. There is very little margin for error and variation in the chemical components of the cell. Unless the chemistry of the cells of an organism is strictly regulated, then it soon dies and degrades. That regulation is itself dependent on other amazingly complex chemical systems and is controlled by the DNA and RNA in the cells. The problem for evolutionary theorists is to try and conceive of ways that all this could be gradually built up from less complex systems, and without any intelligent input from outside the systems themselves.

The most common idea is that small organic molecules gradually accumulated in an organic soup, and that these then reacted with one another to produce more complex ones. Somehow, these became partially isolated from their surrounding environment and a new stage of chemical evolution began, leading eventually to self-replicating single cells that in turn became many-celled organisms. The experiments of Miller in the 1950s were once thought to be a hopeful way of creating new organic molecules. He passed electric currents through a mixture of simple inorganic molecules—thought to simulate the conditions that might have existed on earth billions of years ago. His work produced a variety of new, slightly more complex molecules. However, the results he obtained do not really begin to explain how the much larger organic molecules that are vital to life were formed. It is also questionable whether the conditions he envisaged as pertaining to the supposed early earth ever really existed.

These difficulties in formulating mechanisms by which life could have arisen without intelligent design or action have been looked at by a number of different mathematicians. Indeed a whole conference was devoted to it in the 1980s. Their conclusions were clear. The mathematical odds against such a chance origin of life are so great that it was deemed to be impossible. Those who argue that it must be possible, since life is here, are merely stating their belief against all the odds and evidence. The odds of the same person winning the lottery four times in succession look like a sure bet in comparison!

It is such considerations that have led some, like Fred Hoyle, to postulate that life did not originate on earth at all, but was brought to earth on meteorites. He was not talking about aliens in the popular sense of that word, but some simple form of life from which has developed all the more complex forms. He came to this conclusion because he was convinced that the possibility of life arising here by chance was so small as to be impossible. He still believed in life arising without divine intervention, but he pushed the problem into outer space.

Is there life out there?

This brings us to an important question. Is there any evidence of life existing in this vast universe apart from on the earth? This is certainly a matter of great interest to many scientists and others. Vast resources have been devoted to the hunt for extra-terrestrial life forms. Attempts range from experiments carried by the machines exploring nearby planets looking for chemical signs of life, to the search for intelligent communications from far-distant civilisations. For some decades scientists have been scanning radio frequencies for any sign of intelligent life trying to communicate with us on earth. It is hoped to discover patterns of radio or other signals that cannot be explained by random 'noise'. So far nothing has been heard at all. Of course, those involved in the search point to the vastness of the universe, and the relatively short time we have been looking for any signs, as the reason why we have heard nothing yet.

A little time ago great excitement arose in some quarters at claims that fossilised signs of life had been found on a Martian meteorite found in Antarctica. The suggestion made was that the rock originated on Mars and had been blasted from its surface through the impact of a meteorite with such violence that it escaped the gravity of Mars and escaped into space. It was then captured by earth's gravity and fell to earth. The 'primitive' life that exists on Mars had been carried as a passenger on this unusual interplanetary transport! The claim that these are fossils of life forms has, however, been rejected by most other scientists, and the 'fossils' attributed to quite normal chemical crystals.

This raises the question of whether some form of life may exist, or has ever existed, on Mars. The present explorations taking place on its surface are seeking to find traces of such life. It needs to be noted, however, that even if any were to be found, that does not mean that life must have arisen on Mars and was later carried to earth to act as the seed upon which evolution got to work. It is far more logical to conclude that life arrived on Mars on meteorites blasted from the surface of earth—which, after all, teems with life—than vice versa. The chances of life arising by chance on the inhospitable surface of Mars are even smaller than those of it arising spontaneously on earth.

We might ask, Why should anyone expect to find signs of life—intelligent or not—in other parts of the solar system, our galaxy or elsewhere? The answer, of course, is a prior belief that life arose by chance on earth and, given similar conditions, it could arise anywhere. The discovery of planets circling distant stars—even if as yet those found seem unsuitable to sustain life—has encouraged this speculation. As there are billions of stars, the likelihood that another planet like earth exists somewhere is, we are told, inescapable. That may even be so—time will perhaps tell. But even given the same physical conditions conducive to life existing on another planet as those that exist on earth, that still doesn't mean that life has arisen there. The same difficulties to life arising by chance exist wherever the planet may be. In other words, the search for extraterrestrial life is really based on a belief that has not been proved.

Since atheistic scientists are just as committed to their belief system as Bible-believing Christians are to theirs, it is inevitable that the search for alien life forms will continue in order to try and find support for their faith. Every discovery of a new extra-solar planet will be greeted with enthusiasm. The discovery of simple organic molecules in other galaxies will no doubt be heralded as proof that life could arise by chance in other parts of the universe.

Of course, anyone who believes the teaching of the Bible believes in extraterrestrial intelligent beings! The frequent mention of spiritual beings, both good and bad, is clearly taught in Scripture. Angels are ministering spirits, who serve God and God's people. Demons are unclean spirits, who are opposed to God and do all they can to destroy his kingdom. Their influence is seen in their tempting of humans to sin, as well as in other ways. One of their most successful strategies is encouraging people not to believe the Word of God, but to substitute its teaching for the teachings of men. Evolution is one of their great successes! Millions are convinced of the theory of evolution and reject the biblical concept of creation. They do what Paul tells us of in Romans chapter 1. They worship the creation rather than the Creator. They ascribe to things made the capability of doing what only God could do. Having denied God, what else can an atheist do? He embraces the lie of the Evil One rather than believe the Truth of God's Word!

10
Does it matter?

ONE of the questions sometimes asked by Christians is, Does it matter whether we are created by God or evolved by chance over millions of years? Some argue that how we got here is irrelevant to our faith and walk with God. Others naively say that they believe in creation *and* evolution; they claim that God may even have used evolution to make us, so both are true. Other Christians are afraid that standing against evolutionary teaching just makes Christians look ridiculous. In this chapter I want to say why I believe the question of evolution or creation is important.

It affects the way we look at the Bible

I have no doubt that there are sincere Bible-believers who believe in evolution. However, I confess that I find it hard to see how they can read millions of years of evolution into the first chapter of the Bible. Although it is sometimes claimed that Genesis chapter 1 shows a gradual development of life through the creation week, the order of creation is certainly not that of evolutionary theory. The first life forms created on Day 3 were the land plants, while evolutionary scenarios always have the first life forms developing in some form of primeval sea. Even more interesting is the fact that the plants were made a day before the sun had taken its present form in the heavens. The sea creatures were created on Day 4, before the land animals on Day 6. That means that whales and other sea mammals would have been in existence before the land mammals had appeared. So in order to make Genesis 1 teach evolution, it is necessary to say that the words do not mean what they seem to say! Various methods of interpreting the chapter have been suggested in order to get round these obvious contradictions.

One suggestion is the Gap Theory, which suggests that there is a 'gap' of millions of years between verses 1 and 2 of Genesis 1. Following the initial creation of the universe in verse 1 it is claimed that the world was destroyed by God after the Fall of Satan in heaven, and that verses 2-31 record the re-creation of the world following the destruction. A passage that is sometimes quoted to support this interpretation is 2 Peter 3:5-7. However, the context of that statement makes it clear that Peter is talking about the Flood at the time of Noah, and not another Flood that had nothing to do with man. Furthermore, Hebrew scholars tell us that the original Hebrew of Genesis does not suggest a re-creation of the earth.

Another suggestion is the so-called Framework hypothesis. This ingenious interpretation says that there is a clear parallelism in Genesis 1. The sea is made on Day 1 and populated on day 4. Likewise, the heavens are created on Day 2 and its inhabitants (the birds) on Day 5. Similarly, the land is made on Day 3 and the animals which live on it on Day 6. This structure is said to be a literary device used by Moses to simply teach us that God made all things. While this may be an interesting interpretation of Genesis 1, it is hard to see that it is clearly taught in that chapter! It is especially difficult to accept that the church of God has been misinterpreting these words of Scripture for thousands of years, and that it is only now that are we finally in a position to understand them!

It affects the way we look at ourselves and God

The Bible teaches us that mankind was made in the image of God. He possesses an eternal soul or spirit that will never cease to exist. He has been made with the capability of having fellowship with God his Creator. Furthermore, he is responsible for his actions and must one day give an account of his life before the throne of God his Judge. Jesus came into the world to save us from the consequences of our rebellion against the laws of God. He died on the cross in our place that we might be forgiven, suffering himself what we so rightly deserve. Those who believe what the Bible teaches are motivated to live their lives in the fear of the Lord and in gratitude for his great mercy and grace.

If we teach our young people that they are the products of a blind, impersonal, meaningless process called evolution, we should not be surprised if they live very differently. If life has no purpose and there is no reckoning and judgement to come, why should I fear what I do, at least so long as I don't get caught? 'Eat, drink and be merry; for tomorrow we die', has little power to constrain sin and selfishness. If there is no God and no judgement to come, what does it matter what an individual does? Who has the right to say what is right or wrong? Indeed, if evolution is true, how can we even talk about right and wrong or good and evil? There is no such thing, morally, as good or evil; the only value judgement we can make is whether a thing is positive or negative for the survival of our species! Indeed, the individual has no real worth in himself, especially if he should be regarded as a hindrance to the evolutionary advance of humanity.

Since we are supposed to have come from a process of death and destruction leading to the emergence of higher and better life forms, why should it be wrong for the strongest in society to dictate what we may or may not do? It is because we have been created that it matters what we do. One day God is going to judge all men, and he has given proof of this by raising Jesus from the dead. Of course, some do not believe that. They do not believe in the resurrection—because the dead are not raised—science has proved it, they tell us! That, however, does not alter the fact that the resurrection of Jesus is a fact of history. As created sinful beings, it is time to seek the Lord! I can close this chapter with no better words than those of Ecclesiastes:

Remember your Creator in the days of your youth . . . before the silver cord is loosed . . . Then the dust will return to the earth as it was, and the spirit will return to God who gave it . . . For God will bring every work into judgment, including every secret thing, whether it is good or whether it is evil.

(Ecclesiastes 12:1,6,7,14)

In the final chapter we shall look at the Bible's picture of history and the future.

11
The Bible's story

WE live in a wonderful, beautiful, complex world and universe. At whatever level we look at it, whether the microscopic world of the cell or the immense cosmos of the galaxies and space, we see the order and design of God the Creator: 'The heavens declare the glory of God, and the firmament shows his handiwork' (Psalm 19:1).

That order is clearly seen in the opening chapter of the Bible, where we find the account of God's creative activity. Over the six twenty-four-hour days of creation this universe was brought into being by the Word of the Lord. God spoke and it was done. A perfect world appeared step by step. Following the appearance of the various kinds of plants and animals that God made, small gradual changes within the borders of created kinds were immediately possible. No new genetic information was needed for changes to occur, simply the sorting out of the already created genetic variability within each kind. Adam and Eve were placed in a perfect world.

A perfect creation has changed
However, our experience and observation of life tells us that the world is far from perfect today. Disease and sickness, famine and floods, destructive storms and tornadoes, earthquakes, parasites and dangerous animals, as well as the fact of death—all are now a part of the everyday experience of the inhabitants of this world. The question that

is frequently and understandably asked is, Why is that so, if God made all things and God is so good? I believe that it is only as we take seriously the Bible's account of the fall of man into sinful rebellion against the law of his Creator that we can begin to make sense of this paradox.

God made Adam and Eve in his own likeness, with a moral, spiritual nature. They were put into the world to look after it and steward it for God's honour and glory. Fellowship with their Creator was both their privilege and pleasure and their normal experience before they sinned. However, the act of placing their judgement of what should or should not be done before that of God's brought untold misery into the world. They disobeyed the clear and simple command of God. The third chapter of Genesis describes what happened on that fateful day in their lives. They chose to be independent of God. No longer willing subjects of their Maker, they desired autonomy and self-determination rather than obedience and trust.

They got far more than expected! Their own natures were corrupted and changed in consequence and they became sinful in their being. It became normal for them and their descendants to choose selfishness and sin instead of holiness and goodness. However, in the midst of the sadness and awfulness of man's ingratitude and folly, God gave the promise of One who would come to destroy the work of the devil (Genesis 3:15).

The effect of man's sin on the creation
Fellowship with God was broken by their sin, and his holy anger against unrighteousness and evil was shown against them both. They were driven out of the paradise he had created for them, and at the same time the rest of the universe was cursed by God. Principles of disorder and decay

were deliberately introduced by the Creator into his creation. In the words of the apostle Paul,

> The creation was subjected to futility, not willingly, but because of him who subjected it in hope; because the creation itself also will be delivered from the bondage of corruption into the glorious liberty of the children of God. For we know that the whole creation groans and labours with birth pangs together until now.
>
> (Romans 8:20-22)

How God effected the changes we do not know, but he who made all things would certainly know how to change the underlying physical laws of nature to accomplish his purposes. The principle of death began to operate in both man's experience and the created order. Life that had been easy because of the abundance of God's provision in the Garden of Eden suddenly became more difficult and wearisome:

> Cursed is the ground for your sake; in toil you shall eat from it all the days of your life. Both thorns and thistles it shall bring forth for you, and you shall eat the herb of the field. In the sweat of your face you shall eat bread till you return to the ground, for out of it you were taken; for dust you are, and to dust you shall return.
>
> (Genesis 3:17-19)

In the following chapters of Genesis we see that the sentence promised for disobedience was carried out: after each name we read, 'and he died' (Genesis 5). The world was now different from how God had originally created it. Some animals may at this stage have become carnivores instead of vegetarians. Other differences from the world today were certainly present, though we cannot say exactly what they were, since God has not revealed them to us.

The descendants of Adam and Eve possessed great inborn capabilities and intelligence, and it is no surprise to read of the rapid development of culture and technology in the early centuries of the world as the human population grew rapidly. The discovery of music, metalworking, the building of towns and the development of agriculture are mentioned.

At the same time, the environment of the early earth was conducive to life. This, together with the absence at first of disease and parasites, which would have gradually come into existence, coupled with the almost perfect genetic make-up of the early populations until mutations appeared, led to great ages among mankind. These factors may also account for the gigantic forms of plants and animals that we see in the fossil record.

Judgement fell upon a corrupt world

The sinful tendencies of humans continued, and violence and corruption came to dominate the world to such an extent that something had to be done. God judged the world through the Flood of Noah. The account of that judgement in Genesis chapters 6–8 makes for sombre reading. A holy God will not tolerate sin, and to safeguard the promise of a Saviour who would be a descendant of Eve (the promise already given in the Garden of Eden), God started again with Noah, who found grace in his eyes, and his family. All other human beings perished. He who gives us life is also the one who takes it away!

Billions of animals were also wiped out in the Flood. Their fossilised remains testify to the universal worldwide nature of the catastrophe. Some species within the kinds probably became extinct at this time, but the kinds continued. Only the representatives of each animal kind saved in the Ark emerged at the end of the year-long judgement to continue to populate the now much-changed earth. As they multiplied and met different environments as they slowly dispersed over the earth, the inherent genetic variability in each kind of plant and animal was again subjected to selection; new species emerged within the kinds, and others died out.

A message of hope

God gave his promise to Noah, however, and through him to mankind, that he would never again destroy the earth through a flood. The Bible continues then to tell the story of how the Lord God chose Abraham and his descendants to belong to him, and how the promise of the coming Saviour was renewed and clarified through the centuries. The focus of the promise, announced by Israel's prophets, was narrowed down, first to the tribe of Judah, then to the family of David, king of Israel.

Eventually, in the fullness of time, God's purpose in human history reached a point where it was time for the promise to be kept. Jesus was born of the virgin Mary in Bethlehem. The Creator of the universe, the eternal Son of God, took upon himself human flesh and became man! He came to deal with the problem of our sinfulness. He lived a life we could not live—a life of perfect obedience to God. After thirty-three years he offered himself up to God as a substitute for sinful people. He died on the cross as a sacrifice for our sins. He died in our place, taking our punishment upon himself, that we through his death might be forgiven and be reconciled to God. God showed his acceptance of that offering by raising him from the dead. Jesus lives and promises that all who trust in him as Saviour receive pardon for sin and everlasting life. Through faith in his name we can have fellowship with God our Creator again!

However, the effects of Christ's death on the cross are far greater even than the salvation of an innumerable company of people who would believe on him! It affects the whole of creation. He is the Saviour of the cosmos too. One day he is going to return from heaven, to which he ascended after his resurrection, and then not only is he going to receive his people to himself so that they can dwell with him for ever, but he is also going to restore all things to how they were intended to be!

He will cleanse this present universe through fire and create a new heavens and a new earth, in which dwell righteousness. It will be a universe without death, disease, devil or danger, one that will show the power, perfection and purity of the everlasting, loving, holy, wise God our Saviour, and one that will be a suitable home for God's people for eternity. Those who live in it will lack nothing and will spend their days—if 'days' is an appropriate word to use for that blessed existence—in enjoying God and the work of his hands for ever.

> Therefore, beloved, looking forward to these things, be diligent to be found by him in peace, without spot and blameless; and reckon that the long-suffering of our Lord is salvation.
>
> (2 Peter 3:14,15)

If this is not yet your hope, then call upon the name of the Lord that you may be saved!

Bibliography

There are many good books available today dealing with aspects of the Creation–Evolution controversy from a biblical viewpoint. The following are some that I have found helpful even when I didn't agree with all their conclusions.

General

Bone of Contention, Sylvia Baker (Evangelical Press, 1986).

Creation & Change: Genesis 1:1–2:4 in the Light of Changing Scientific Paradigms, Douglas Kelly (Christian Focus, July 2008).

Did God Use Evolution? Werner Gitt (New Leaf, 2006).

Genesis for Today, Andy McIntosh (DayOne, 1997).

Science vs Evolution, Malcolm Bowden (Sovereign Publications, 1991).

Scientific Creationism, Henry M. Morris (Creation Life Publishers, 1985).

The Evidence for Creation, McClean, Oakland & McClean (Whitaker House, 1989).

What About Origins? Monty White (Dunestone, England, 1978).

The Age of the Universe

How Old Is the Earth? Monty White (Evangelical Press, 1985).

It's a Young Earth After All, Paul Ackerman (Baker Book House).

Origin and Destiny of the Earth's Magnetic Field, T. G. Barnes (Institute for Creation Research, 1983).

Starlight and Time, D. Russell Humphreys (Master Books, 1994).

The Age of the Solar System, H. S. Slusher & S. Duursma (Institute for Creation Research, 1978).

The Early Earth, J. C. Whitcomb (Evangelical Press, 1972).

The Young Earth, John D. Morris (Master Books).

Thousands Not Millions, edited by Donald de Young (Master Books, 2005).

Fossils

The Fossils Still Say No! Duane Gish (Master Books, 1995).

Bones of Contention: A Creationist Assessment of Human Fossils, M. Lubenow (Baker Books, 1992).

In the Beginning: Compelling Evidence for Creation and the Flood, W. Brown (Centre for Scientific Creationism, 1996).

Ape-Men: Fact or Fallacy? M. Bowden (Sovereign Publications, 1988).

Dinosaurs: Those Terrible Lizards, Duane Gish (Creation Life, 1978).

The Great Dinosaur Mystery Solved, Ken Ham (New Leaf Press, 2000).

Genetics

Genetic Entropy and the Mystery of the Genome, John Stanford (Elim Publishing, 2005).

After Its Kind, Byron Nelson (Bethany Fellowship, 1980).

* * *

I would also like to recommend some Websites, where answers to many questions can be found, as well as articles dealing with a number of specific scientific and biblical topics.

Answers in Genesis: www.answersingenesis.org

Biblical Creation Society: www.biblicalcreation.org.uk

Creation Science Movement: www.csm.org.uk

Truth in Science: www.truthinscience.org.uk